MICROCHIP

SMALL WONDER

MICROCHIP
SMALL WONDER

CHARLENE W. BILLINGS

Illustrated with photographs

A Skylight Book

DODD, MEAD & COMPANY
New York

Frontispiece: Courtesy Intel Corporation

In memory of my mother

Acknowledgments

My sincere appreciation to everyone who has helped to provide information and photographs for this book. Special thanks to my husband, Barry, who has given technical guidance throughout the preparation of the manuscript.

Library of Congress Cataloging in Publication Data

Billings, Charlene W.
  Microchip : small wonder.

  (A Skylight book)
  Includes index.
  Summary: Explains what a microchip is, what it is used for, how it is made, and how it works.
  1. Microelectronics—Juvenile literature. [1. Microelectronics] I. Title.
TK7874.B55   1984   001.64   84-10179
ISBN 0-396-08452-4

CONTENTS

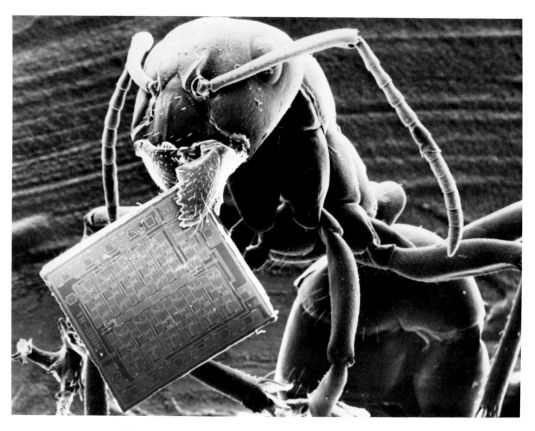

Taken with a Philips Scanning Electron Microscope, this photograph shows an ant holding a memory storage chip from a computer.

North American Philips Corporation

1

WHAT
IS A CHIP?

Consider an object that you can hold on the tip of your finger. It is so light you could blow it away with a puff of your breath. It is so small an ant could carry it off.

Yet this object will make your life far different than your mother's or father's life. This object is one of the most important inventions of recent years. And though you may not be aware of it, it already is all around you in things you see or use every day.

What is this small wonder—this tiny object that is changing our world?

It is called the chip.

Chip is a nickname for *microchip*. *Micro* means very

small. Scientists and engineers who work with micro-chips shorten the term to chip. By either name, the microchip has become part of all our lives.

At first glance, a chip could be mistaken for a fleck of gray paint. But if you look more carefully, you notice that it is a perfect square, about as thick as a thumbnail.

Close-up of a microcomputer chip

Courtesy of Texas Instruments, Inc.

Each side of the tiny square is only one-quarter of an inch long.

There is a pattern on the surface of the chip. But it is too small to see with the naked eye. You need a very strong magnifying glass or microscope to see the design clearly.

If the pattern is made large enough to see, it looks like the map of a city. Row upon row of "streets" overlap and intersect. These "streets" are really circuit lines that serve as wires to carry electricity through the chip.

Besides circuit lines, there are thousands of very, very tiny electronic parts built into the chip. The circuit lines and electronic parts guide and control electricity as it passes through the chip.

The microchip, then, is a network of miniature circuit lines and electronic parts. This tiny device is amazing because it can take the place of thousands of much larger wires and parts. But just as important are that the chip costs far less and can do more work—better and faster—than anything before it could do.

WHERE
ARE CHIPS USED?

The invention of the microchip has brought about many vast changes in only a few years.

The shrinking size and cost of computers are examples of this. The first electronic computers were built in the early 1940s. One of these computers was called ENIAC (which stands for Electronic Numerical Integrator and Calculator). It was a thirty-by-fifty-foot giant that filled a room as big as a school gymnasium. ENIAC weighed over thirty tons! It was built with more than 18,000 glass electronic tubes and miles of wires that took over two and one-half years to put together. A computer like ENIAC cost millions of dollars. Not even big businesses wanted to spend that much for a computer.

ENIAC, one of the first electronic computers

Courtesy of Apple Computer, Inc.

Miss H. L. Marvin at keyboard of an early-day complex computer. Photograph taken in 1940.

Courtesy of Bell Laboratories

Modern computers made with microchips are dwarfs by comparison. They fit onto desktops or are small enough to be easily carried. Nevertheless, they can do more work and do it faster than ENIAC. And they cost far less, some only a few hundred dollars.

Early computers like ENIAC were built with thousands of fragile glass electronic tubes. Each tube took about as much space as a skinny light bulb. The tubes used a lot of electricity. They were costly to run and would get hot when the computer was in use. Almost every time these computers were used, some of the tubes would overheat and break down. The wires connecting the tubes and parts to each other inside the computer sometimes would come loose. Engineers could spend

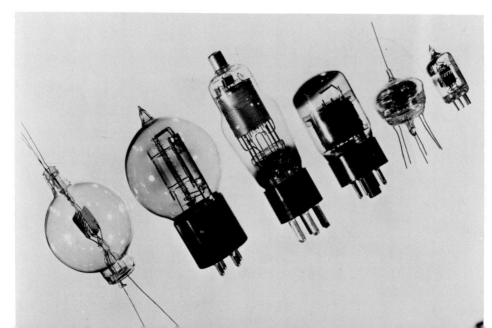

Courtesy of Bell Laboratories

hours looking for the faulty connection or tube. Meanwhile, the computer could not be used.

The first breakthrough in solving these problems of early computers came in 1947 with the invention of the *transistor*.

Transistors could do the same jobs as the glass electronic tubes. That is, they could make a weak electrical signal strong. They could control the flow of electricity through a computer. And, they could act as a switch that could be ON or OFF.

Compared to the glass electronic tubes, the first transistors were small—only about the size of a pencil eraser. They also used a lot less electricity. They did not get hot and were much less expensive to run.

Still, each transistor had to be wired into place individually. So even though the first transistors were a great improvement, engineers wanted to make computers yet smaller, more reliable, and less costly to buy and use.

By the early 1960s scientists had found ways to make much smaller transistors and circuit lines on a single sliver of material. These tiny chips could do the same

Early glass electronic tubes or vacuum tubes

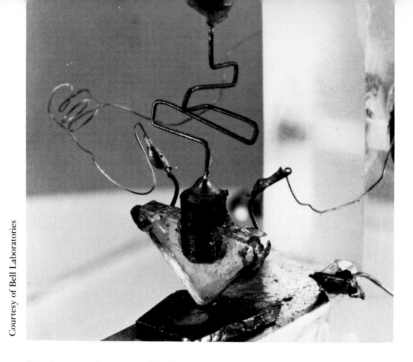

The first transistors assembled by their inventors at Bell Laboratories in 1947 were primitive by today's standards.

jobs as the bulky glass electronic tubes or the individually wired transistors of the older computers. The microchip was a reality!

The microchip uses very small amounts of electricity. It does not get hot and costs little to run. Because no wires can come loose within the chip, a computer built with microchips is far less likely to fail.

Thus, the chip has made it possible to reduce the size and cost of computers. Now many people can own a

An early computer built with transistors instead of vacuum tubes.

Courtesy of Bell Laboratories

home computer. Many schools also are able to provide computers for their students.

Chips are the "brains" in today's computers. But there are many other things around us that are made with microchips.

Handheld electronic calculators owe their smallness to the chip. You can do more mathematics in two minutes with one of these calculators than in half an hour with paper and pencil. In 1970, a palm-sized calculator that could add, subtract, multiply, and divide cost over four hundred dollars. It used so much power that its batteries would last only about ten hours.

Nowadays a calculator that can do the same things costs under ten dollars. It may be no bigger than a few playing cards stacked together and will run on one battery the size of a shirt button for a year or more.

Another item that uses chips is a microwave oven. Many of these ovens have touch panels instead of knobs and dials to turn. Behind these panels are microchips. The cook can give a series of instructions to the oven by touching the pads on the panel. The microchips store the instructions and then automatically control all of the

An array of articles which use microchips

steps to do the cooking. The cook need only return when the food is ready.

Did you know that you may be wearing microchips on your wrist? Digital watches were unknown until a few years ago. Wind-up watches with dial-like faces were the only kind of watches available. A digital watch displays the time in blocklike numbers that change every minute or second of the day. Many of these watches also have built-in calendars and alarms. Some even include built-in calculators.

In a digital watch, microchips take the place of the

Top: An eight-digit memory calculator watch by Timex

Bottom: A computer backgammon game

Ruder Finn & Rotman

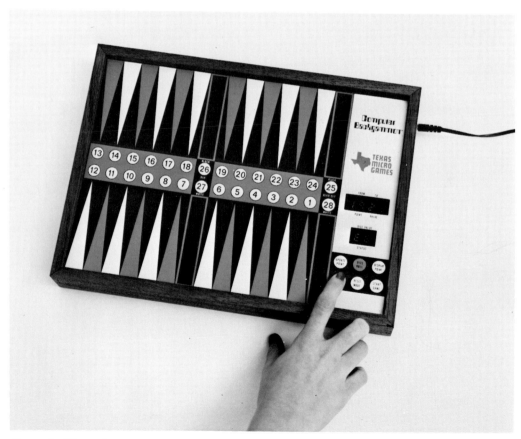

Courtesy Intel Corporation

many little gears of the wind-up watch. The chips not only help to tell the time, they also control the display of the numbers on the face of the watch.

Digital watches will run for a year, using small batteries. And they are known for their accuracy. At the end of a full year of keeping time, digital watches will still be within a second or two of the exact time. Digital watches can be bought for no more than the cost of a wind-up watch.

When you think of robots, do you think of microchips as well? Chips control the actions of robots that are at work in factories in the United States and other countries. Many of these robots do jobs such as spray paint or carry heavy items for people. They also weld, assemble, polish, shape, mix, or cut materials. Robots are useful to do hazardous tasks too. In the future robots may fight fires or mine minerals from deep beneath the sea for us.

A robot even is being developed to explore other planets. It will be able to collect rocks and other samples as it moves along the surface. On its own, it will avoid craters, cliffs, and other dangers.

Where else are microchips in use? When you play a video game you are using microchips. Hidden inside the game are chips that keep track of the time and the score. Other chips store things like the rules and the sequence of events for the game. Many electronic games can fit into a pocket—thanks again to the tiny chip. If you have played these games, you know that a lot can happen in a blink of time. Some of these games even have sounds such as a victory tune when you win—or a raspberrylike "*BLAT*" when you lose.

The wires of a telephone can bring the news, weather, scores of sports events, stock market reports, and much other information to someone who can connect their home computer to their telephone. The information is processed by the microchips inside the computer and quickly appears on the computer's televisionlike screen.

Microchips also help in medicine. The chips housed inside computers can store descriptions of thousands of illnesses and their treatments—more than doctors could possibly remember by themselves. A doctor can feed a patient's symptoms into a computer and get back a diagnosis and treatment. A storage bank of information like this is especially valuable to care for rare diseases that are not often seen by doctors.

Handicapped people are being aided by the chip as well. Devices small enough to be placed permanently under the skin or scalp can bring back some sight or hearing for blind or deaf people. Better artificial limbs with more lifelike control are available for those who have lost arms or legs to disease or accident.

Besides earthbound uses, the exploration of space would be impossible without the microchip. The United States sent the *Viking* spacecrafts to Mars. *Voyager 1* and *2* flew by Jupiter and Saturn. The crafts sent back information and close-up photographs of these planets unlike any ever seen by humans before. The shuttle *Columbia* has flown into space several times and returned to earth to be used again. Lightweight computers made with microchips controlled every phase of these missions.

Chips are in things as different as cash registers and heart pacemakers. They control the emissions from cars, traffic lights on street corners, thermostats in buildings, and the newest household appliances. Chips are inside radios, television sets, and automatic cameras as well.

There are many more uses for microchips. Most were hardly imagined only a few years ago.

The space shuttle Columbia

HOW
MICROCHIPS ARE MADE

If you think about the tiny size of the chip—and the extraordinary things it can do—it is hard to believe that people can make such a thing.

What materials do engineers use to design and build this minimarvel? How are hundreds of thousands of circuit lines and transistors placed in a perfect pattern onto a chip that can pass through the eye of a needle?

Most microchips are made from a substance called silicon. Silicon is very plentiful in the earth's crust. It is the main ingredient in sand. The silicon used to make chips usually comes from quartz rock.

Electricity flows easily through some materials such as

copper, gold, aluminum, and other metals. Such materials are called *conductors*. Electricity cannot pass through materials such as rubber, plastic, and glass. These are known as *insulators*.

Silicon is neither a good conductor of electricity nor a good insulator. It is somewhere in between. For this reason silicon is called a *semiconductor*.

However, what is important is that silicon can be treated and changed to make it serve as either a good conductor of electricity *or* a good insulator. This is why

A chip can pass through the eye of a needle.

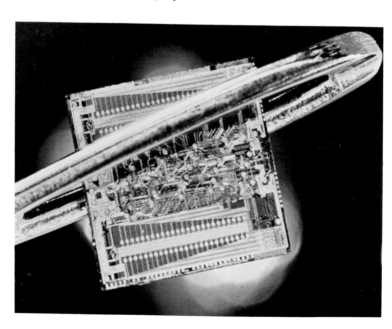

Courtesy of International Business Machines Corporation

silicon is one of the most useful materials for making microchips.

Silicon is melted and then rehardened into a single, solid crystal three to four inches across and up to several feet long. The hard crystal is ground into a smooth rod shape.

Rod-shaped crystals of silicon from which wafers are cut.

Courtesy of American Microsystems, Inc. (AMI) Santa Clara, California

A high-speed diamond saw is used to cut the large rod of silicon into thin wafers—much like cutting a loaf of bread into slices. Each delicate, blue-gray slice of silicon is no thicker than a few pages of this book.

One side of each wafer is polished to be perfectly flat and smooth, and shiny as a mirror.

The silicon wafers are heated in a special high-temper-

Technician carrying a tray of wafers

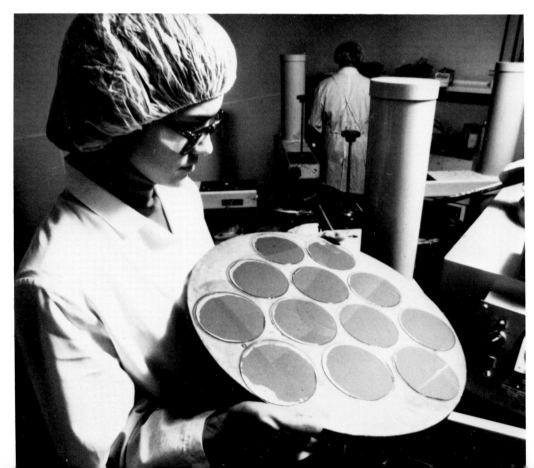

Courtesy of American Microsystems, Inc. (AMI) Santa Clara, California

ature oven with oxygen. During this treatment, oxygen combines with silicon on the polished surface of each wafer. A very thin layer of silicon dioxide or glass forms. This layer acts as an insulator—electricity cannot flow through it.

Once this insulating layer is formed, each wafer is ready to have as many as one hundred or more identical microchips created on its surface. When completed, each chip will have many layers of interconnected circuit lines and transistors.

Because just one speck of dust could ruin some of the chips while they are being made, wafers must be processed into chips in special "clean rooms." The air in these rooms is filtered to keep out the tiniest particles of dirt. People who work in these rooms wear jumpsuits or coats and caps made from lint-free fabric.

Engineers draw a design for each layer of the microchip. Often, they use computers to aid them. Then photographs of the designs are reduced to the tiny size of the chip. The final size is hundreds of times smaller than the drawings first made by the engineers.

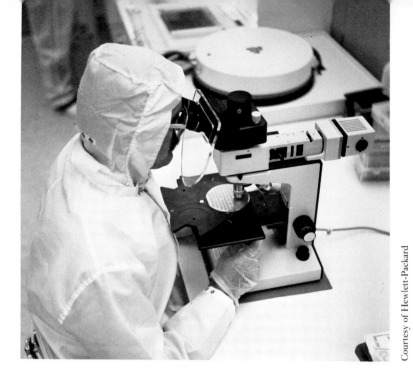

Courtesy of Hewlett-Packard

Operators in "clean rooms" wear special clothing.
Circuit for a microchip is laid out with the aid of a computer.

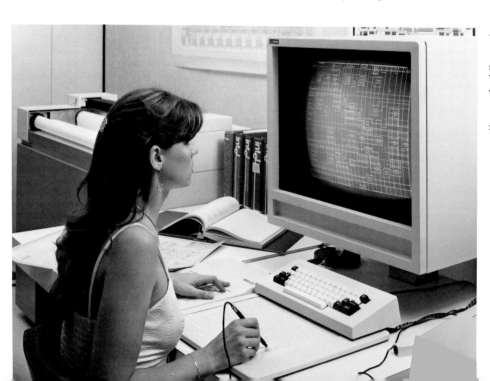

Courtesy Intel Corporation

The design for a layer on the chip is transferred onto a glass sheet called a *mask*. The design is repeated in columns and rows on the mask so that many identical chips can be made on a wafer at one time.

The stencil-like mask is placed over a wafer that has been coated with a special film. Then light is passed through the mask onto the wafer.

Wherever light passes through the mask, the coating on the wafer becomes hardened. But the coating hidden by the designs on the mask stays soft.

The wafer is rinsed with chemicals that wash away the soft portions of the coating. Another chemical removes the thin insulating layer just beneath. The areas of silicon exposed in this way are exact copies of the design on the mask for each chip.

Before continuing, the hardened filmlike coating on each wafer is stripped away with strong chemicals.

Now the exposed areas of silicon are treated with substances called *dopants*. The dopants change the silicon in these areas so that it conducts electricity.

Each wafer is coated with a very thin new layer of

silicon by heating it with silane—a gas that contains silicon.

By repeating similar steps with different masks, several layers of patterned semiconductor, insulator, and conductor are formed on each chip. Incredibly small, overlapping areas of the differently treated silicon create the thousands of transistors on each chip.

Fine circuit lines of aluminum or gold are added to connect the transistors of each microchip to small metal pads along the edges of each chip.

After the microchips on a wafer are finished, they are checked to make sure that they work properly. Needle-like probes, so small they are watched under a microscope, touch and test each chip on a wafer.

With testing completed, each wafer is separated into as many as one hundred or more microchips.

Finally, each microchip is encased in a mounting. Sturdy metal prongs outside of the mounting are connected by fine wires to the small metal pads along the edges of the chip inside. Electricity flows through the prongs and fine wires into and out of the chip. The prongs make the chip easy to plug in.

By 1971, engineers had produced a microchip that included the main parts of a computer all on one piece of silicon. A "computer-on-a-chip" is called a *microprocessor*. These chips are found inside most home computers.

A diamond saw is used to cut the wafers into chips.

Researchers also use beams of electrons or X-rays to "write" circuit lines and transistors directly onto chips. These newer methods allow scientists to make the circuit lines of chips much thinner than ever before. This opens the way to microchips that can do even more.

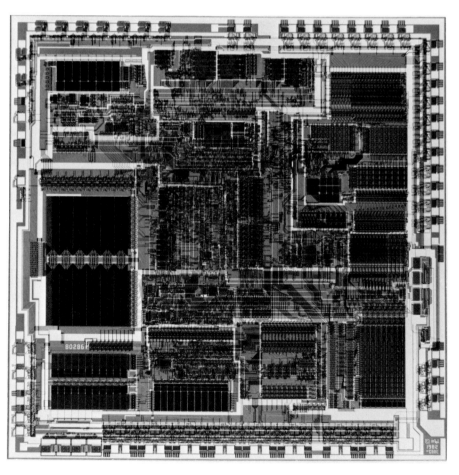

HOW
MICROCHIPS WORK

When the pattern on a microchip is made larger, it looks like the map of a city. In a city, the flow of cars and trucks along the streets is regulated by traffic lights.

In a chip, the "traffic" is made up of electrical signals that travel in pulses along the "streets" or circuit lines. Instead of traffic lights, tiny transistors control the flow of signals.

But, you still wonder, how can people use a microchip as a tool to count and calculate? How do chips handle words and store information for us?

One of the things a transistor can do is act as a switch. And a switch can be either ON or OFF.

A microprocessor by Intel Corporation

An employee checks designs for a finished microchip circuit.

You use switches every day when you turn electric lights on and off in your home. When the switch is on, the light is on. When the switch is off, the light is off. You know instantly which way the switch is by whether the light is on or off.

The thousands of transistors built into microchips that act as switches make it possible for a computer to count and calculate for us.

Computers count electrically. To a computer a switch that is ON stands for a 1. A switch that is OFF stands for a zero or nothing. Because only two digits or numbers are used, this is called a *binary system*. The word "binary" means two. Each zero or 1 is called a *binary digit* or *bit*.

To count like a computer, imagine a counter made of four light bulbs arranged in a row. Each of these four light bulbs has a number over the bulb. Many computers use the numbers 8, 4, 2, and 1. You will soon understand why.

Each light bulb has a switch to turn it ON or OFF. When a bulb is ON, it stands for the number above it. When a bulb is OFF, it stands for zero or nothing.

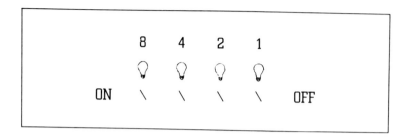

When all of the switches are OFF, all of the light bulbs are OFF. The computer counts zero.

To count 1, the switch under light bulb 1 is turned ON. The light bulb turns on and the computer counts 1.

To count 2, the switch under light bulb 1 is turned OFF. The switch under light bulb 2 is turned ON. The light bulb that stands for 2 comes on. The computer counts 2.

To count 3, the switch under light bulb 1 is turned ON again. Now both the 2 and the 1 light bulbs are on. The computer adds these together to get 3.

As you can see from the diagrams of light bulbs and switches, by using different combinations of the numbers 8, 4, 2, and 1, only four switches are needed for the computer to count up to 15.

A Four-switch Counter Made with Four Light Bulbs

	8	4	2	1	
ON	💡	💡	💡	💡	OFF
	\	\	\	\	
	8	4	2	1	
0	0	0	0	0	
1	0	0	0	■	
2	0	0	■	0	
3	0	0	■	■	
4	0	■	0	0	
5	0	■	0	■	
6	0	■	■	0	
7	0	■	■	■	
8	■	0	0	0	
9	■	0	0	■	
10	■	0	■	0	
11	■	0	■	■	
12	■	■	0	0	
13	■	■	0	■	
14	■	■	■	0	
15	■	■	■	■	

0 = Light bulb is OFF   ■ = Light bulb is ON

By using more four-switch counters, computers can count to enormous numbers.

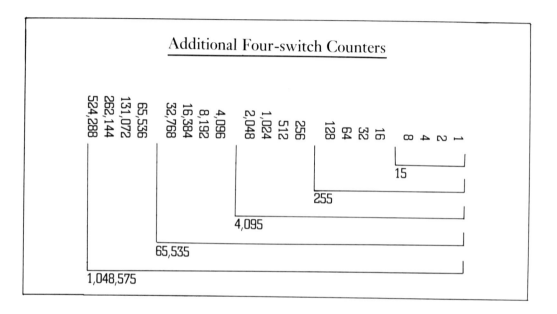

Additional Four-switch Counters

The value of the numbers of each additional four-switch counter continues to double. Thus, with only twenty switches, a computer can count to over one million.

To get a better grasp of how the binary system works, try using your fingers to count like a computer.

First write the number 1 on the nail of the middle finger of your right hand and the number 2 on the nail of the index finger of the same hand. Now write a 4 on the nail of the index finger and an 8 on the nail of the middle finger of your left hand.

Place your hands palms down in front of you on a flat surface such as a table or desktop. Form fists with your hands. Then extend only the marked fingers of each hand.

To count you will need to follow this rule: A finger that is up stands for the number marked on it. A finger that is down stands for a zero.

To count zero, all four fingers are down or touching the table.

To count 1 put the finger marked 1 up.

To count 2, put the finger marked 1 down and the finger marked 2 up.

To count 3, put the finger marked 1 up again. Add the 2 and the 1 together to get 3.

Now can you figure out which fingers of the left and right hands are up and which are down for the number 4

. . . the number 5 . . . all of the numbers from 6 to 15? Compare your answers to the diagram of light bulbs to see if you are correct.

The way computers count may seem awkward to you. But a computer made with microchips can count faster electrically than humans can think. The electrical signals that pass through the circuit lines and transistors of a chip travel at nearly the speed of light. Each switch can go ON or OFF more than one *million* times in less than one second! This is why a computer made with micro-chips can add, subtract, multiply, and divide thousands of huge numbers as fast as a flash.

To understand how a microchip works with words, think again of a switch that can be either ON or OFF. Now think of these switches as a way to send word messages using a code. The code has only two symbols— ON and OFF. You have seen that ON and OFF can stand for numbers. But they can stand for words too.

In the 1830s a man named Samuel Morse invented the telegraph to send messages quickly over long distances. The telegraph uses a code with only two symbols. Morse Code uses combinations of DOTS and DASHES to stand for any letter of the alphabet, any number, and even for punctuation such as periods and question marks.

In a similar way, combinations of ON or OFF switches can be used as a code for words or information. Thus, each ON or OFF switch stands for one piece or bit of information. Usually, eight of these bits of information are grouped together to form one *byte*. A microchip can handle in an instant the coded bits and bytes of a message that would take a telegraph operator many minutes to send in Morse Code.

Here are two codes that both use only two symbols to send messages or store information. One is the Morse Code and the other is known as the American Standard Code for Information Interchange or ASCII-8.

Character	Morse Code	ASCII-8
a	. —	11100001
b	— . . .	11100010
c	— . — .	11100011
d	— . .	11100100
e	.	11100101
f	. . — .	11100110
g	— — .	11100111

h	. . . .	11101000
i	. .	11101001
j	. – – –	11101010
k	– . –	11101011
l	. – . .	11101100
m	– –	11101101
n	– .	11101110
o	– – –	11101111
p	. – – .	11110000
q	– – . –	11110001
r	. – .	11110010
s	. . .	11110011
t	–	11110100
u	. . –	11110101
v	. . . –	11110110
w	. – –	11110111
x	– . . –	11111000
y	– . – –	11111001
z	– – . .	11111010
. (period)	. – . – . –	01001110
?	. – . . – .	01011111

Time how long it takes you to tap out the alphabet in
Morse Code with your fingertip. Can you do it from a to

z in one minute? A microchip can do the same thing thousands of times in only one second!

How can microchips also store information for people? Some chips are designed for this purpose. They are called *memory chips*. Memory chips are important because they can store very large amounts of information. Groups of switches built into these chips are set at either ON or OFF. They stay as they are set. In this way the switches "hold on to" or store information.

Scientists measure how many bits of information are stored on a chip in K's. Each K (from the word kilo that means 1,000) is just over 1,000 bits of information. Thus, a microchip that has a 256K memory can store more than 256,000 bits of information.

An array of memory components

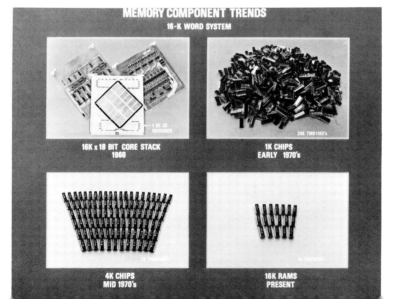

Other chips are called *logic chips*. The switches built into these chips can process information, do arithmetic, and make decisions.

The tiny chip is awe-inspiring for all of the reasons mentioned so far. But a microchip is a device made by people. It cannot do anything on its own. For example, someone must give directions to the microchips within a computer that tell it exactly what to do. These step-by-step instructions are called a *program*. The program can be entered into the chips of a computer by typing it on the computer keyboard. Later, the instructions in a program can be changed to suit a new problem or task.

People can use microchips to solve an almost endless variety of problems or do any number of jobs. Therefore, they are one of the most important inventions of our time.

A logic chip for an electronic calculator, size 1.4" × .6"

Courtesy of Texas Instruments, Inc.

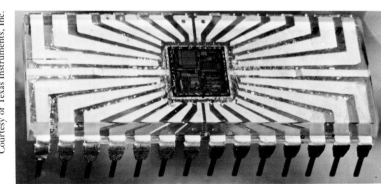

5

MICROCHIPS
OF THE FUTURE

What is in the future?

People who design microchips are still trying to squeeze more and more circuit lines and transistors onto each sliver of silicon. By creating chips that can do more, engineers will make possible new computers that will be easier for people to use. These "user-friendly" computers will be instructed to recognize human speech and respond to spoken commands. They will even be able to tell one person's voice from another.

Scientists will be able to store much more information inside the microchips of the future. Computers made with these chips will be able to do things that today's computers cannot do. Tomorrow's computers will be able to translate back and forth between languages such as Japanese (that uses thousands of difficult characters) and English (that uses the alphabet).

Some scientists are working on new kinds of electronic switches. One kind operates at very high speeds when kept extremely cold. These switches can change from OFF to ON or back again more than a *billion* times in one second. Chips made with these switches could bring about supercomputers that carry out 60 million instructions each second. This is ten times faster than today's best computers.

Researchers also are working to create what they call a *biochip*. In a biochip, new kinds of man-made molecules will serve as electronic switches and circuits. What is more extraordinary is that scientists will be able to instruct these molecules to assemble more molecules like themselves.

The biochip seems like science fiction. But it may be no more impossible in the future than today's microchips would have seemed thirty years ago.

Microchips have already brought about astonishing changes. In the years to come, chips will be more powerful tools than ever—helping us in ways we can scarcely imagine now. People will use microchips to increase their knowledge and improve their lives.

The stamp of progress. Steady progress in complexity and power marks the history of the microprocessor.

Courtesy of Bell Laboratories

INDEX